NASA's SpaceX Crew-9 Adventure

Exploring the Cosmos Everything You Need to Know About NASA's SpaceX Crew-9

James J. Smith

All rights reserved. No part of this publication may be reproduced, distributed, or transmitted in any form or by any means, including photocopying, recording, or other electronic or mechanical methods, without the prior written permission of the publisher, except in the case of brief quotations embodied in critical reviews and certain other noncommercial uses permitted by copyright law.

Copyright © **James J. Smith** 2024

Table of contents

Chapter 1	**6**
Introduction to NASA's SpaceX Crew-9 Mission	**6**
Overview of the Crew-9 mission objectives	15
Importance of the mission in the context of space exploration	24
Chapter 2	**27**
Crew Members of NASA's SpaceX Crew-9 Profiles and Their roles	**27**
Chapter 3	**38**
Mission Details and Timeline	**38**
Launch preparations and schedule for the SpaceX Crew-9 mission	43
Description of the Dragon spacecraft and Falcon 9 rocket	48
Orbital parameters and docking procedures with the International Space Station (ISS)	53
Chapter 4	**57**
Scientific Objectives and Research Activities	**57**
Overview of the scientific experiments and research conducted during the mission	62
Contributions to various fields of study, including biology, biotechnology, physical science, and Earth science	67
Chapter 5	**72**
Beyond Low Earth Orbit: Future of Space Exploration	**72**

Discussion on the significance of the ISS as a testbed for long-duration spaceflight   **77**

Exploration of commercial opportunities in low Earth orbit and NASA's Artemis campaign for lunar exploration   **82**

Conclusion   **88**

Intentionally left blank

Chapter 1

Introduction to NASA's SpaceX Crew-9 Mission

Have you ever felt an overwhelming need to know more as you gazed up at the starry sky, a canvas of shimmering moonlight and stars, and a sense of wonder wash over you? Has the immensity of space ever sparked a desire to test our limits and discover what else is out there? If so, then you are invited to join NASA's SpaceX Crew-9 mission for an exciting adventure.

This journey isn't like any other circumnavigation of the planet. It's a huge step forward in human exploration, proving that we've been eager to learn more about the world ever since we looked into the first campfire flames. The Commercial Crew Program, an innovative program that has changed the face of space travel, is celebrating its ninth crewed trip with Crew-9. Picture this: Launched from American territory, rockets and spacecraft manufactured in the United States are escorting a crew of astronauts from around the world on a quest to improve human knowledge.

An Orchestra of Experience: The Fearless Crew

A true trailblazer, Commander Zena Cardman is at the head of this daring undertaking. This is her inaugural expedition to the universe, a culmination of years committed to revealing the treasures lying beneath the Earth's surface. Her experience in geobiology adds a fresh viewpoint to the team, as she studies the links between life on Earth and the potential for life beyond.

Veteran astronaut Nick Hague adds a wealth of experience to the mission. His previous mission in 2018 may not have gone as planned, witnessing a near-tragic launch abort, but his tenacity and unwavering drive to space exploration are a monument to the human spirit. As the pilot, his calm temperament and refined talents will be key in maneuvering the Crew Dragon spacecraft through the immense distance.

Stephanie Wilson, a seasoned astronaut with three past space shuttle missions under her belt, represents the unwavering spirit of discovery. As Mission Specialist 1, she'll leverage her significant experience in operating the robotic arm, a critical tool for conducting research and maintaining the International Space Station (ISS). Her knowledge in engineering and aeronautical science will be vital in assuring the smooth running of the expedition.

Rounding out the crew is Aleksandr Gorbunov, a cosmonaut from Roscosmos, Russia's space agency. This is his first spaceflight, a monument to the multinational partnership that has become a hallmark of space exploration. His engineering skills, acquired during his employment with

Rocket Space Corporation Energia, will be important in assuring the success of the mission.

A Technological Marvel: The Crew Dragon and Falcon 9

Imagine a sleek, white capsule, resembling a futuristic dragonfly, preparing for liftoff. This is the Crew Dragon, a wonder of modern engineering designed by SpaceX. This state-of-the-art spacecraft will be the crew's home for six months, giving them with the life support, communication, and maneuvering capabilities necessary to accomplish their mission. Equipped with touchscreen controls, automated docking systems, and advanced life-support technologies, the Crew Dragon represents the culmination of human intellect and aspiration.

But the Crew Dragon doesn't begin on this voyage alone. The massive Falcon 9 Block 5 rocket serves as its flaming chariot, a pillar of flame and power that will push the spaceship on its astounding trip. Standing 230 feet tall, this reusable launch vehicle epitomizes the next phase of space travel, giving a more cost-effective and sustainable approach to space exploration.

Destination: The International Space Station - A Collaborative Hub for Humanity

The crew's final goal is a wonder of international cooperation – the International Space Station (ISS). Imagine a gigantic building, a tribute to human ingenuity, orbiting the Earth at 17,500 miles per hour. This man-made satellite has been continually occupied since 2000, acting as a platform for groundbreaking scientific inquiry, technological innovation, and international collaboration. The group-9 team will become temporary residents of this stunning orbiting station, joining a multinational group dedicated to pushing the boundaries of human knowledge.

A Six-Month Odyssey: Unraveling the Mysteries of Space

For six months, the Crew-9 crew will become pioneers on the cutting edge of scientific discovery. Their days will be filled with a range of activities, from conducting innovative studies in biology, physics, and materials science, to maintaining the complex systems of the ISS and stepping outside on risky spacewalks to execute crucial repairs and upgrades.

Imagine cultivating plants in microgravity, potentially paving the path for future space farms that could sustain long-duration missions. Picture astronauts investigating the effects of space radiation on the human body, critical knowledge for future journeys to Mars. Think of them testing novel materials in the harsh environment of space, pushing the boundaries of what's possible for future spacecraft and habitats.

But our mission isn't simply about scientific discovery and technical advancement. It's about the human spirit — the unquenchable curiosity that motivates us to seek the unknown, to push the boundaries of what's possible. Imagine the awe-inspiring view from the cupola, a panoramic window providing astronauts a spectacular sight of our home planet, a swirling blue marble floating in the blackness of space.

Think of the weightlessness, the freedom of mobility, the ability to float unrestrained within the spacecraft. Picture the friendship that will bloom among this foreign crew, as they rely on each other for survival and support in the hostile environment of space. This project is a monument to the power of human collaboration, a beacon of hope in a world often divided.

The Road Ahead: A Stepping Stone to the Future

The journey of Crew-9 is significantly more than just a six-month expedition. It's a stepping stone on the route to a future where space exploration is no longer the province of science fiction, but a realistic reality. As we inch closer to our goals of exploring beyond Earth, Crew-9 sets the stage for future expeditions to the Moon and maybe Mars.

The data obtained, the knowledge gained, and the international collaboration established will be invaluable in defining the future of space exploration. This mission is an inspiration to a new generation of scientists, engineers, and visionaries, a tribute to the unlimited potential of humanity.

Join the Adventure: A Book for All

This book is your invitation to embark on this remarkable voyage alongside Crew-9. Within these pages, you'll experience the excitement of launch, the wonder of space exploration, and the breakthrough science being undertaken aboard the ISS. Whether you're a seasoned space fan or merely interested about the world beyond our atmosphere, this book has something for you.

We'll delve into the complexity of spaceflight, the challenges and triumphs of living in microgravity, and the cutting-edge science that is defining the future of space exploration. You'll meet the crew members, each with their own distinct stories and motives, and experience the international partnership that makes this trip possible.

This book is not only about the destination; it's about the journey itself. It's about the voracious human curiosity, the everlasting spirit of adventure, and the potential for revolutionary discoveries that lay just beyond the horizon. So, buckle up, my reader, and prepare to be sent on a fantastic voyage onboard NASA's SpaceX Crew-9 mission. As Carl Sagan famously stated, "Somewhere, something incredible is waiting to be known." Let's go on this voyage together and discover the wonders that await us a million miles above.

Overview of the Crew-9 mission objectives

Our exciting voyage aboard NASA's SpaceX Crew-9 mission continues! We've met the intrepid crew, marveled at the technological marvel that is the Crew Dragon, and recognized the significance of the International Space Station (ISS) as a collaborative hub for humanity. Now, it's time to go deeper into the core of this mission: the scientific objectives that will move us forward in our quest to understand the universe.

Imagine the Crew-9 crew as explorers exploring into new land, each experiment an important step on the route to groundbreaking discoveries. Their mission objectives include a wide range of scientific fields, each contributing to a broader knowledge of our role in the cosmos.

Unveiling the Mysteries of Life: Biology Takes Center Stage

Biology reigns supreme on Crew-9's agenda. Imagine astronauts producing plants under weightless circumstances, nurturing small seedlings into brilliant greens. This isn't simple gardening; it's an important experiment that could pave the way for future space farms. Think of astronauts meticulously researching the behavior of microscopic creatures on the ISS, observing how these tiny living forms adapt to the harsh environment of space.

The knowledge obtained from these tests could alter our understanding of life beyond Earth. It could potentially unveil the secrets of how creatures survive and thrive in severe settings, paving the path for the detection of life on other worlds.

But the focus isn't simply on alien life; Crew-9 will be exploring the impacts of spaceflight on human biology. Imagine astronauts partaking in rigorous exams, measuring their bone density, muscular mass, and cardiovascular health. These investigations hold the key to comprehending the challenges of long-duration space travel, allowing us to create effective countermeasures that can safeguard the health of future astronauts journeying deeper into space.

Pushing the Boundaries: Physics Experiments Take Flight

Think beyond biology; Crew-9 goes into the interesting area of physics. Imagine astronauts doing experiments that challenge our concept of gravity, examining how fluids behave and objects interact in the microgravity environment of the ISS. The data acquired could alter our knowledge of fundamental forces and pave the way for the development of new technologies on Earth and beyond.

Furthermore, Crew-9 will study the exciting domain of fluid dynamics. Picture astronauts investigating the behavior of liquids in microgravity, observing how they flow, mix, and react in this unique environment. This understanding could be crucial in the design of future spaceship propulsion systems and life-support technology.

Building a Better Future: Materials Science Takes the Stage

Imagine astronauts evaluating novel materials aboard the ISS, subjecting them to the harsh environment of space. These experiments hold significant relevance for the future of space travel. The data acquired could pave the way for the development of stronger, lighter, and more durable materials for future spacecraft, habitats, and even spacesuits.

Think of Crew-9 testing materials specifically developed to withstand the intense radiation bombardment encountered in outer space. These advances could be vital for shielding humans on voyages to the Moon and Mars, insulating them from the hazardous effects of cosmic radiation.

Beyond the Experiments: The Importance of Daily Operations

The Crew-9 mission isn't just about innovative research; it's about maintaining the key systems that keep the ISS running. Imagine astronauts tirelessly doing normal maintenance jobs, guaranteeing the seamless running of the station's life-support systems, power grids, and communication equipment. Their dedication offers a safe haven for research and a platform for future exploration.

Spacewalks: exploring beyond the ISS

A significant component of Crew-9's mission entails traveling outside the protective shell of the ISS. Picture astronauts embarking on daring spacewalks, linked to the station by lifelines, as they execute crucial maintenance and upgrading operations. Imagine them methodically upgrading broken equipment, installing new scientific experiments, and doing repairs on the station's façade.

These spacewalks are not only crucial for the continuous operation of the ISS, but they also give invaluable experience for future trips to the Moon and Mars. They allow astronauts to refine their skills in functioning in a vacuum, moving with cumbersome spacesuits, and communicating efficiently under duress.

Connecting the Dots: How Crew-9 Fits into the Bigger Picture

The objectives of Crew-9 are not standalone activities; they're important threads woven into the wider tapestry of space travel. The data collected, the experience gained, and the technologies tested all contribute to a greater purpose - pushing the boundaries of human knowledge and driving us further into the universe.

Think of Crew-9 as a bridge between the present and the future. The scientific achievements obtained onboard the ISS will shape the design and development of future spacecraft, pave the way for human missions to the Moon and Mars, and unlock new opportunities for space exploration in the years to come.

A Shared Journey: International Collaboration on the ISS

The Crew-9 mission isn't only about scientific discovery; it's a monument to the strength of worldwide collaboration. Imagine the crew members, hailing from the United States and Russia, working together flawlessly aboard the ISS. This spirit of cooperation is a vital component of space exploration, requiring mutual respect, good communication, and a shared desire for pushing the boundaries of human achievement.

The success of Crew-9 rests on the crew's ability to operate together as a cohesive one. They come from various backgrounds and cultures, but they share a single purpose — to increase scientific knowledge and explore the wide unknown. Through this collaboration, they set an example for future generations, showcasing the potential for international cooperation to achieve great things.

The Crew-9 Mission: A Beacon of Inspiration for Future Generations

Imagine a little youngster staring up at the stars, their eyes full with wonder and curiosity. Crew-9 acts as a beacon of encouragement for this child and countless more like them. This mission proves that the seemingly impossible can be done through devotion, hard effort, and a thirst for knowledge.

By following the crew's voyage through the pages of this book, young readers can get a deeper understanding of space exploration, the scientific method, and the necessity of worldwide collaboration. They may very well be encouraged to pursue jobs in science, technology, engineering, and mathematics (STEM) sectors, paving the way for a future generation of space explorers.

A Million Miles Above - More Than Just a Mission

As we go deeper into the objectives of Crew-9, it becomes evident that this mission is significantly more than just a collection of scientific experiments. It's a tribute to the enduring human spirit, a symbol of worldwide cooperation, and an inspiration for future generations.

By following the crew's trip aboard the ISS, we get a look into the future of space exploration, a future when mankind pushes the boundaries of knowledge, journeys further into the cosmos, and unlocks the enormous potential that awaits us a million miles above.

Importance of the mission in the context of space exploration

The value of the Crew-9 mission may be recognized through numerous essential components that contribute to the wider picture of space exploration:

Scientific Advancement:

Pioneering Biology Research: Crew-9's investigations on plant growth and creature behavior in microgravity pave the way for future space farms and unveil secrets of life beyond Earth. Understanding how organisms adapt to severe settings could be vital for finding life on distant worlds.
Unlocking the Effects of Spaceflight: Studying the crew's health under microgravity helps us create countermeasures for long-duration space travel, ensuring astronaut health for future trips to the Moon and Mars.

Pushing the Boundaries of Physics and Materials Science: Crew-9's studies on fluid dynamics and material testing lead to breakthroughs in spaceship propulsion systems, life-support technologies, and stronger materials for future spacecraft and habitats.

Technological Innovation:

Crew Dragon and Falcon 9: The mission utilizes a reusable spacecraft and launch vehicle, suggesting a more sustainable and cost-effective approach to future space exploration missions.

International Collaboration:

Symbol of Unity: The multinational crew aboard the ISS displays the potential of international cooperation, developing mutual respect and communication, necessary for future large-scale space endeavors.

Inspiring the Next Generation:

A Beacon of Hope: Crew-9's mission acts as an inspiration for young minds, kindling a desire for STEM areas and paving the path for a future generation of space explorers.

Stepping Stone to the Future:

Laying the Groundwork: The data, experience, and technology tested on Crew-9 are vital for future missions to the Moon and Mars, altering the trajectory of space exploration for years to come.

In essence, Crew-9 is more than a solo mission; it's a stepping stone on the route to a future when space travel is more accessible, sustainable, and filled with new discoveries. It's a monument to human curiosity, teamwork, and our continuous drive to understand the world beyond our globe.

Chapter 2

Crew Members of NASA's SpaceX Crew-9 Profiles and Their roles

Zena Cardman: Commander - A Pioneer at the Helm

Hailing from Williamsburg, Virginia, Zena Cardman is more than simply a mission commander; she's a pioneer with a remarkable past. Growing up, she was attracted by the natural world, spending her childhood exploring the woodlands and marshes of Virginia, her curiosity aroused by the intricate ecosystems hiding beneath the surface. This curiosity drove her to obtain a bachelor's degree in Biology and a master's degree in Marine Sciences at the University of North Carolina at Chapel Hill.

Zena's specialty specializes on geobiology, a field that examines the interaction of life with geological processes. This unique perspective is essential to Crew-9, as she will be entrusted with conducting tests meant to study how life adapts to severe circumstances, maybe offering clues about the existence of life on other worlds. Imagine her methodically cultivating plants under microgravity, analyzing their growth patterns and responses to the harsh space environment. This data could be the key to deciphering the secrets of extraterrestrial life.

However, Zena's leadership qualities transcend beyond scientific interest. She has a solid track record in team management and crisis response, honed during her years as a research team leader at NASA's Ames Research Center. Her calm attitude, effective communication skills, and ability to inspire confidence will be key in leading the diverse Crew-9 team through the challenges and successes of their six-month mission on the ISS. Picture her on the bridge of the Crew Dragon, issuing directions with a serene authority, her eyes expressing the vast joy of space travel.

Nick Hague: Pilot - Resilience Forged in Fire

Nick Hague is a monument to the unshakeable human spirit. His previous launch attempt in 2018 was a terrible near-tragic incident, encountering a launch abort just seconds after liftoff due to a rocket booster failure. Despite this setback, Nick's ambition to shoot for the stars remained undimmed. He spent the following months undertaking rigorous training and examinations, displaying his unwavering commitment to space exploration.

Nick offers a wealth of experience to Crew-9, having already logged over 203 days in space during his previous mission aboard the International Space Station in 2019. His calm temperament under pressure and his polished piloting talents are significant assets to the crew. Imagine him at the helm of the Crew Dragon spacecraft, his attention concentrated on the instrument panel, directing the convoluted trajectory towards the ISS with steady precision. Nick is more than simply a pilot; he's a symbol of resilience, perseverance, and the unbreakable human spirit that propels us to explore the unknown.

Stephanie Wilson: Mission Specialist 1 - A Master of the Robotic Arm

A seasoned astronaut with three earlier space shuttle missions under her belt, Stephanie Wilson represents the everlasting spirit of discovery. Possessing a natural gift for engineering and a penchant for problem-solving, Stephanie flourished throughout her earlier missions, doing innovative research and participating in important spacewalks.

However, Stephanie's main competence lies in controlling the Canadarm2, the robotic arm of the ISS. This complicated mechanical marvel is a critical instrument for conducting experiments, deploying satellites, and performing maintenance on the exterior of the space station. Imagine Stephanie delicately guiding the Canadarm2 with the precision of a surgeon, her moves precise and efficient as she undertakes delicate duties outside the station. Her experience and technical talents will be vital for the success of Crew-9's research activities.

Aleksandr Gorbunov: Mission Specialist 2 - A Bridge Between Nations

Aleksandr Gorbunov, a cosmonaut from Roscosmos, Russia's space agency, gives an international viewpoint to Crew-9. His selection for this mission illustrates the ongoing spirit of teamwork that has become a hallmark of space exploration. Aleksandr is not just an astronaut; he's a bridge between nations, a symbol of the unifying human ambition to explore the wide cosmos.

With a strong foundation in engineering, refined during his employment with Rocket Space Corporation Energia, Aleksandr boasts a profound understanding of spacecraft systems and maintenance processes. Imagine him diligently troubleshooting technological issues onboard the ISS, his skill assuring the seamless operation of the station's complicated systems. Aleksandr's maiden spaceflight is more than just a personal success; it's a monument to the strength of international cooperation, opening the way for a future when nations work together to push the boundaries of human achievement.

These four individuals, each with their own unique backgrounds and skillsets, embody the best of what humanity has to offer. Together, they comprise the intrepid crew of Crew-9, a monument to human curiosity, collaboration, and the persistent will to explore the unknown. Their six-month mission onboard the ISS promises to be a journey of startling discoveries, stretching the boundaries of scientific knowledge and ...inspiring a new generation of dreamers and innovators. As you go deeper into this book, you'll get to know each crew member on a personal level. We'll explore their motives for pursuing professions in space exploration, the hurdles they overcome during their training, and the personal sacrifices they made to go for the stars. You'll witness their camaraderie grow as they navigate the limited confines of the ISS, relying on each other for support and motivation through the ups and downs of their journey.

A Day in the Life of Crew-9: A Glimpse into the Extraordinary

Imagine waking up to the beautiful sunrise above Earth, a swirling blue marble poised in the blackness of space. This is the crew's daily reality aboard the ISS. But their days are far from leisurely. Their agenda is methodically organized, complete with a tough blend of scientific inquiry, station maintenance, and physical activity to offset the effects of microgravity.

The Thrill of Experimentation: Unveiling the Mysteries of Space

A considerable amount of their day is dedicated to undertaking groundbreaking experiments. Picture Zena diligently analyzing plant development patterns, while Stephanie skillfully steers the Canadarm2 to install a fresh science package. Aleksandr might be assisting Nick with calibrating instruments for a physics experiment meant to examine the behavior of fluids in microgravity. These seemingly routine tasks contain great significance, each one a tiny step on the journey to discovering the universe's secrets.

Maintaining a Home in Space: Ensuring the Smooth Operation of the ISS

The crew isn't simply doing experiments; they're also responsible for maintaining the key systems that make the ISS running. Imagine them meticulously replacing air filters, checking life-support systems, and conducting normal maintenance duties, ensuring a safe refuge for research and a platform for future exploration.

Spacewalks: Venturing Beyond the ISS

A highlight of the journey for some crew members will be heading outside on risky spacewalks. Imagine them linked to the station by lifelines, their heavy spacesuits sheltering them from the harsh elements of space. They'll be tasked with crucial maintenance tasks, such as installing new equipment or repairing failing systems. These spacewalks not only contribute to the continuous operation of the ISS, but they also give essential experience for future missions to the Moon and Mars.

Leisure Time: A Glimpse of Human Connection in the Cosmos

Despite their demanding schedules, the crew also finds time for leisure activities. Imagine them meeting in the Cupola, a panoramic window affording stunning views of Earth, sharing stories, and communicating with loved ones back home via video chats. These moments of connection serve as a reminder of the human spirit that propels them to explore the immense unknown, even as they hurtle through space at 17,500 miles per hour.

Challenges and Triumphs: The Inevitable Ups and Downs of Spaceflight

Living and working in a limited area for six months will provide its own set of obstacles. Imagine the crew coping with emotions of loneliness, the ever-present sense of being millions of kilometers from home, and the potential for technical faults or unexpected problems. However, these trials will also develop a spirit of solidarity between the crew members. They will learn to rely on one other for support, share achievements together, and overcome difficulties as a united team.

A Beacon of Inspiration: Crew-9's Legacy

The journey of Crew-9 is considerably more than just a scientific mission; it's a monument to the undying human spirit. Their purpose acts as a beacon of inspiration for a new generation of dreamers and innovators. By following their tale within the pages of this book, young readers will develop a deeper understanding of space travel, the scientific method, and the value of international collaboration. Who knows, this book might just spark a desire for STEM areas in a young mind, paving the way for a future generation of astronauts and space explorers.

As Crew-9 begins on their unprecedented six-month journey, one thing is certain: they are not simply astronauts; they are pioneers, pushing the boundaries of human knowledge and leaving their mark on the history of space travel. Join their trip within the pages of this book, and ready to be amazed.

Chapter 3

Mission Details and Timeline

Launch: The Crew-9 mission is slated for a launch no sooner than August 2024 atop a SpaceX Falcon 9 rocket from Launch Complex 39A at NASA's Kennedy Space Center in Florida. This timeframe is subject to change and will be determined closer to the launch date.

Crew:

Commander: Zena Cardman (USA) - Geobiologist
Pilot: Nick Hague (USA) - Veteran Astronaut Mission Specialist 1: Stephanie Wilson (USA) - Robotics Expert Mission Specialist 2: Aleksandr Gorbunov (Russia) - Engineer

Mission Duration: The Crew-9 mission is intended to endure for approximately six months. During this time, the crew will live and operate aboard the International Space Station (ISS), doing scientific research, maintaining the station's equipment, and engaging in spacewalks.

Mission Phases:

1. Launch and Docking: The mission begins with the launch of the Crew Dragon spacecraft on a Falcon 9 rocket. Following a successful launch, the Crew Dragon will autonomously maneuver and dock with the ISS, typically within 24 hours.
2. Station Integration and Science: After docking, the Crew-9 astronauts will undertake a brief adaptation period to adjust to the microgravity environment. Once adapted, they will shift to a demanding schedule loaded with scientific research across numerous areas including biology, physics, materials science, and human health studies.
3. Spacewalks: Throughout the mission, there will be a defined period for Extravehicular Activities (EVAs), also known as spacewalks.

During these excursions, one or two crew members will walk outside the ISS, tethered for safety, to undertake important maintenance work, upgrade experiments, or conduct repairs on the station's exterior.

4. Mid-Mission Resupply: During the six-month mission, one or two cargo resupply trips are routinely scheduled. These missions, delivered by spacecraft like SpaceX's Dragon or Northrop Grumman's Cygnus, will provide the crew with crucial supplies, equipment, and even fresh meals.

5. Undocking and Re-entry: As the mission nears its completion, the Crew Dragon will undock from the ISS and commence its descent back to Earth. During re-entry, the spacecraft will encounter scorching heat before deploying parachutes and splashing down softly in the Atlantic Ocean off the coast of Florida, where recovery teams will be waiting to aid the crew.

Benefits of the Mission:

Scientific Advancement: Crew-9's study adds to a broader understanding of living in space, human health in microgravity, and the creation of new technologies.

worldwide Collaboration: The multinational crew illustrates the strength of worldwide cooperation in space exploration.

Inspiration for the Next Generation: The mission acts as a beacon of inspiration for young minds, stimulating an interest in STEM subjects and future space exploration activities.

Following the Mission:

The Crew-9 mission will be thoroughly documented by NASA and its international partners. You can expect to track the mission's progress through many means, including:

NASA TV: Live broadcasts of launch, docking, and undocking events.

NASA Website: Mission updates, crew profiles, and educational tools.

Social Media: Follow NASA and the crew members on social media sites for real-time updates, images, and videos.

Launch preparations and schedule for the SpaceX Crew-9 mission

As of today, April 22, 2024, there is no official notification from NASA or SpaceX regarding the particular launch preparations and schedule for the Crew-9 mission. However, based on existing facts, we can make some reasonable guesses:

Target Launch Window: We know the launch is slated for no sooner than August 2024. This timescale is susceptible to vary and could be altered by several events, including:

Technical Readiness: Both the Crew Dragon spacecraft and the Falcon 9 rocket need to undergo comprehensive testing and inspections to assure mission success. Any unanticipated technological concerns could cause delays.

Range Availability: Launch windows are meticulously planned to assure airspace safety and best circumstances for launch and recovery. Other launches or other activities can demand alterations to the Crew-9 launch window.

Weather Conditions: Favorable weather conditions are vital for a safe launch. Unforeseen weather patterns could cause a delay.

Possible Launch Preparations:

Crew Training: The Crew-9 astronauts are likely undertaking extensive training at present. This training comprises numerous topics, including: Familiarization with Crew Dragon spacecraft systems and procedures.

Training for diverse mission circumstances, like crises or equipment breakdowns.

Practicing scientific experiments they will do onboard the ISS.

Spacecraft and Rocket Preparations: SpaceX engineers are presumably performing final assembly and testing of the Crew Dragon spacecraft.

The Falcon 9 rocket will also undergo extensive inspections and test firings to assure peak performance.

Launch Site Preparations: Launch Complex 39A at Kennedy Space Center in Florida will be prepped for the flight. This includes preparing ground support equipment, launch towers, and recovery personnel.

Following Launch Preparations:

Pre-Launch Briefings: A few days before launch, NASA and SpaceX will likely have press briefings to discuss the mission details, crew preparedness, and weather outlook.

Static Fire Test: SpaceX might undertake a static fire test of the Falcon 9 rocket on the launchpad. This test ignites the engines for a limited period to assess their operation.

Launch Countdown: Closer to launch day, a comprehensive countdown will be issued, covering the final hours leading up to liftoff. This countdown will involve actions including crew boarding the ship, hatch closure, and rocket fuelling.

To stay updated on the official launch preparations and timeline for Crew-9, you can check the following resources:

NASA website: https://www.nasa.gov/
- Look for detailed updates on the Crew-9 mission under the "Missions" or "Space Station" sections.

NASA TV: https://www.nasa.gov/nasatv/ - They might broadcast live events connected to Crew-9 training or pre-launch briefings.

SpaceX website: https://www.spacex.com/ - Check for news announcements or updates regarding Crew-9 launch preparations.

Social media: Follow NASA and SpaceX on social media sites like Twitter for real-time information and announcements.

Remember, all information is based on current knowledge and the launch timetable can vary. Stay tuned to official channels for the most up-to-date details on the Crew-9 mission.

Description of the Dragon spacecraft and Falcon 9 rocket

The Dragon Spacecraft: A Reusable Haven in Space

The Crew Dragon spacecraft, created and built by SpaceX, acts as the dependable chariot for astronauts on Crew-9. Here's a rundown of its primary features:

Design: The Dragon consists of two basic sections: a reusable capsule and an expendable trunk.
The pressurized capsule provides a pleasant living and working environment for up to seven crew members. Imagine a high-tech cabin equipped with touchscreen controls, comfy seating, and enormous windows affording spectacular views of Earth.
The trunk, positioned beneath the capsule, carries important cargo for the mission, including as supplies, experiments, and docking

devices. Once its contents are emptied, the trunk breaks from the capsule and burns up harmlessly in Earth's atmosphere.

Capabilities: The Dragon is built for reusability, a vital component of SpaceX's aim for cost-effective space travel. The capsule is fitted with heat shields that protect it during the fiery re-entry into Earth's atmosphere. Upon re-entry, parachutes deploy to slow down the drop before a graceful splashdown in the ocean. Recovery crews will then be on hand to collect the capsule and its crew.

The Dragon sports a powerful Draco Thruster System, a network of tiny rocket engines used for maneuvering in orbit, critical for docking with the ISS and conducting necessary course adjustments.

For crucial launch abort scenarios, the Dragon is equipped with a strong SuperDraco Launch Escape System. These powerful engines can quickly drive the capsule away from a faulty rocket in event of an emergency during launch.

The Falcon 9 Rocket: A Powerful Workhorse

The Falcon 9 rocket acts as the muscular workhorse responsible for driving the Crew Dragon spacecraft towards the heavens. Here's a closer look at this fantastic launch vehicle:

Design: The Falcon 9 is a two-stage rocket, meaning it has two independent stages that fire in sequence to achieve orbit. The first stage, the workhorse, is a strong cylinder-shaped structure holding nine Merlin engines, which burn a combination of liquid oxygen and highly refined kerosene fuel. Imagine the flaming boom of these engines as they ignite, generating immense thrust to hurl the entire spacecraft upward.

The second stage is a smaller, sleeker structure that takes over after the first stage separates and falls back to Earth (typically recovered by SpaceX for future flights). This higher stage houses a single Merlin engine, responsible for the final push needed to achieve orbital velocity.

Capabilities: Similar to the Dragon capsule, the Falcon 9 is largely reusable. The first stage is meant to return to Earth after launch, completing a controlled descent and landing either on a drone ship in the ocean or on a designated landing site at Cape Canaveral. This reusability greatly cuts launch costs compared to standard expendable rockets.

The Falcon 9 is a reliable launch vehicle, with a high success rate in transporting payloads to orbit. Its powerful engines and efficient design make Dragon a workhorse for SpaceX, flying not just passenger capsules but also satellites and cargo resupply flights to the ISS.

Working in Tandem: A Match Made in Space

The Dragon spacecraft and the Falcon 9 rocket constitute a strong combination, operating in perfect unison to bring astronauts and cargo to space. The Dragon, with its spacious cabin and enhanced safety systems, provides a secure environment for the crew. The Falcon 9, with its sturdy construction and reusability, offers a strong and cost-effective launch. Together, they represent the bleeding edge of space exploration technology, setting the way for a future where space flight is more accessible and humanity's reach stretches further into the universe.

Orbital parameters and docking procedures with the International Space Station (ISS)

Orbital Parameters: Charting the Crew-9's Celestial Course

The Crew-9 mission will see the Dragon spacecraft launched into a specified low Earth orbit to meet and dock with the International Space Station (ISS). Here's a breakdown of the important orbital parameters involved:

Altitude: The ISS maintains an average altitude of approximately 400 kilometers (250 miles) above Earth's surface. The Crew Dragon will be launched into a comparable orbit to achieve a close approach to the space station.
Inclination: The ISS circles Earth at an inclination of 51.6 degrees relative to the equator. This inclination allows the space station

to overfly most of the Earth's populous areas during its roughly 93-minute orbit. The Crew Dragon's launch trajectory will be carefully tailored to match the inclination of the ISS's orbit.

Orbital pace: Both the Crew Dragon and the ISS will be traveling at a phenomenal pace of around 7.66 kilometers per second (17,500 miles per hour) to maintain their orbit around Earth. Docking procedures will require carefully positioning the Crew Dragon to match the speed and direction of the ISS, providing a seamless and safe docking.

Docking Procedures: A Delicate Dance in Space

Docking with the ISS is a complex and delicate procedure that demands painstaking planning and flawless execution. Here's a simplified breakdown of the essential processes involved:

1. Launch and Initial Orbit: The Falcon 9 rocket launches the Crew Dragon spacecraft into a particular low Earth orbit.

2. Orbit Maneuvers: Throughout the opening stages of the mission, the Crew Dragon will undertake a series of thruster firings to fine-tune its orbit, progressively inching closer to the ISS.

3. Rendezvous Phase: As the Crew Dragon reaches within a prescribed distance of the ISS (usually a few kilometers), a series of precision thruster maneuvers are made to match the speed and direction of the space station. Imagine the spacecraft carefully aligning itself with the ISS, like two ships approaching each other on a wide ocean.

4. Docking Interface: The Crew Dragon is equipped with a docking port designed to interface with a comparable port on the ISS. Using a sophisticated guidance system and visual confirmation from the crew, the spacecraft is methodically moved into place.

5. Docking Latching and Seal Formation: Once exactly aligned, the docking mechanism on the Crew Dragon will latch onto the docking port on the ISS. Seals will then activate, establishing an airtight path between the spacecraft and the space station.

Confirmation and Crew Transfer:

Following a successful docking, astronauts onboard the ISS will conduct leak checks to ensure a safe connection. Once all systems are confirmed functioning, the hatch between the Crew Dragon and the ISS will be opened, allowing the Crew-9 astronauts to join their home for the next six months — the International Space Station.

This docking method is a marvel of modern engineering and a monument to the multinational effort that has made the ISS a reality. It constitutes a vital stage in each mission, allowing astronauts to safely reach their orbital laboratory and begin on groundbreaking scientific efforts.

Chapter 4

Scientific Objectives and Research Activities

The Crew-9 mission isn't only about an exciting flight to space; it's a scientific adventure replete with groundbreaking research pursuits. These studies hold great promise to unveil the secrets of the universe, improve life on Earth, and pave the road for future space exploration. Here's a glimpse at some of the projected scientific aims and research activities planned for Crew-9:

1. Biology and Biotechnology: Unveiling the Secrets of Life

Plant Growth in Microgravity: Imagine a science experiment observing how plants respond to the absence of gravity. Crew-9 will nurture numerous plant species, examining their growth patterns, root development, and responsiveness to light in the microgravity environment. This research could unlock secrets about cultivating food sources on long-duration space flights and potentially on future colonies on the Moon or Mars.

Investigating the Effects of Spaceflight on Microorganisms: The harsh environment of space can have tremendous consequences on biological creatures. Crew-9 will undertake research examining how bacteria and other microbes adapt to microgravity. This research is vital for understanding the potential risks astronauts confront during space travel and for devising mitigation techniques to safeguard their health.

2. Human Health and Life Sciences: Safeguarding Astronauts on Long Journeys

Physiological Changes in Microgravity: Living in microgravity can lead to bone density loss, muscle atrophy, and cardiovascular deconditioning. Crew-9 will participate in numerous investigations to study how the human body adjusts to spaceflight. This research will help create countermeasures to protect astronaut health throughout long-duration missions, paving the door for future crewed trips to the Moon and Mars.

The Impact of Space Radiation on Human Cells: Cosmic radiation poses a severe health concern for astronauts during long spaceflights. Crew-9 will engage in research examining the impact of radiation on human cells. This research is vital for creating efficient shielding systems and potential medicines to lessen the adverse effects of radiation on astronauts traveling further into space.

3. Physical Sciences and Materials Science: Pushing the Boundaries of Knowledge

Fluid Dynamics Experiments: Understanding how fluids behave in microgravity is vital for numerous space applications. Crew-9 will conduct experiments investigating fluid movement, mixing, and combustion in the absence of gravity. This research has applications in areas including spacecraft propulsion systems, life-support technology, and even producing new materials with unique features.

Material Testing in Space: The harsh climate of space can damage materials over time. Crew-9 will engage in tests investigating the endurance of novel materials proposed for use in spacecraft and future space habitats. This research is vital for constructing reliable and long-lasting structures for future space exploration missions.

4. Earth Science and Remote Sensing: Gaining a New Perspective of our Home Planet

Earth study Studies: The unusual vantage position of the ISS enables for complete study of Earth. Crew-9 astronauts will undertake investigations on themes like climate change, natural disasters, and agricultural monitoring. This research gives essential data for environmental monitoring and sustainable development activities on Earth.

These are just a few examples of the intriguing scientific pursuits planned for Crew-9. The data and knowledge collected from these investigations will add greatly to our understanding of space, human health, and the universe itself. The Crew-9 mission is a tribute to the power of scientific exploration, pushing the boundaries of knowledge and inspiring future generations to aim for the skies.

Overview of the scientific experiments and research conducted during the mission

The Crew-9 mission promises a varied range of research investigations across several fields. Here's a quick outline to give you a better understanding:

Focus Areas:

Biology & Biotechnology: This topic studies how biological creatures adapt to the microgravity environment. Experiments can involve growing plants, examining microbial behavior, or analyzing the impacts of spaceflight on organisms.
Human Health & Life Sciences: This area stresses astronaut health throughout long-duration missions. Crew-9 will engage in experiments on bone density, muscular mass,

cardiovascular health, and the impact of radiation on human cells.

Physical Sciences & Materials Science: This area pushes the boundaries of physics and material science in microgravity. Experiments can include examining fluid dynamics, combustion patterns, or testing the endurance of novel materials for space uses.

Earth Science & Remote Sensing: The unique perspective from the ISS enables for thorough Earth observation. Crew-9 might perform studies on climate change, natural disasters, or monitor agricultural resources.

Examples of Specific Experiments:

Plant Growth in Microgravity: This experiment could involve raising various plant species and evaluating their growth patterns, root development, and responsiveness to light in the absence of gravity. Imagine a small greenhouse onboard the ISS!

Investigating Spaceflight on Microorganisms: This research can involve researching how bacteria and other microorganisms adapt to microgravity. Understanding how bacteria behave in space is vital for astronaut health and creating closed-loop life support systems.

Physiological Changes in Microgravity: Crew-9 will engage in tests that measure bone density, muscle mass, and cardiovascular health during the mission. This data is crucial for creating countermeasures to keep astronauts healthy on long space trips.

The Impact of Space Radiation on Human Cells: Experiments might involve exposing human cells to simulated space radiation and studying the consequences. This research is crucial for creating preventive measures against radiation exposure during deep space missions.

Fluid Dynamics Experiments: Imagine studying how fluids flow, mix, and even burn differently in microgravity. This discovery has applications in spaceship design, life-support systems, and even material development.

Material Testing in Space: New materials proposed for use in spacecraft and future space habitats will be evaluated for endurance in the harsh space environment. This assures stable and long-lasting structures for future exploratory operations.

Earth Observation Studies: Crew-9 astronauts might take high-resolution images of Earth, monitor weather patterns, or investigate specific places for signals of climate change or natural disasters. This data is vital for environmental monitoring and sustainable development activities on Earth.

Benefits of this Research:

The Crew-9 mission's research activities are not separate projects. The data and insights obtained will contribute greatly to our understanding of:

Life in Space: How living creatures adapt to microgravity circumstances, laying the stage for future space settlement.

Human Health in Space: Developing techniques to maintain astronaut health over long-duration missions, important for deep space exploration.

Space Technology: Advancing fluid dynamics and material science research for enhanced spaceship architecture and life-support systems.

Earth Observation: Gaining useful data about Earth's health, aiding in environmental monitoring and sustainable development activities.

Conclusion:

The Crew-9 mission is a tribute to the power of scientific investigation in space. By undertaking these various experiments, the crew is pushing the boundaries of knowledge and encouraging future generations to aspire for the heavens. Their research prepares the path for a future when humanity's footprint in space expands, driven by scientific curiosity and the quest for new discoveries.

Contributions to various fields of study, including biology, biotechnology, physical science, and Earth science

The Crew-9 mission's scientific agenda is designed to make major contributions across a wide variety of topics. Here's a breakdown of the projected impact on biology, biotechnology, physical science, and Earth science:

Biology:

Understanding Plant development in Microgravity: Experiments on plant development can give light on how animals adapt to low-gravity conditions. This understanding is vital for creating strategies to cultivate food sources on long-duration space missions and potentially for future space settlements. By examining plant responses to light and gravity, biologists might enhance strategies for controlled-environment agriculture (CEA) systems, perhaps benefiting food production even on Earth.

Microbial Behavior in Space: Studying how bacteria and other microorganisms adapt to microgravity has consequences for preserving astronaut health. Microbes can play a significant role in recycling trash and providing clean water in closed-loop life support systems. Understanding their behavior is vital for developing these systems efficiently. Additionally, studies on microbial adaptation can influence methods to minimize the spread of hazardous infections during space travel.

Biotechnology:

Advanced Life Support Systems: Knowledge obtained from Crew-9's tests can lead to the creation of more efficient and reliable life support systems for future space missions. This could incorporate breakthroughs in closed-loop systems that recycle air, water, and waste, creating a tiny biosphere within a spacecraft.

These technologies could have spin-off benefits for applications on Earth, such as disaster relief shelters or sustainable living habitats in adverse climates.

Physical Science:

Fluid Dynamics in Microgravity: Experiments on fluid dynamics in microgravity can lead to advancements in several fields of physical research. This understanding can be employed to improve spacecraft propulsion systems, optimize heat transfer mechanisms for spacecraft architecture, and even lead to the development of novel materials with unique features based on how fluids interact in space.

Earth Science:

Earth Observation and Monitoring: The unique vantage point of the ISS allows Crew-9 to undertake vital Earth observation investigations. Their data can be used to monitor weather trends, track climate change indicators, and assess the impact of natural disasters. This knowledge is vital for environmental monitoring efforts and creating sustainable behaviors on Earth.

Interdisciplinary Impact:

The brilliance of Crew-9's study lies in its transdisciplinary nature. Experiments in one field can have ripple effects that assist others. For example, studying plant growth in space can inspire the development of new biomaterials or contribute to our understanding of plant stress response mechanisms. Similarly, research on fluid dynamics might have benefits in creating more effective water purification systems for closed-loop life support.

Overall Significance:

The Crew-9 mission's scientific contributions go beyond the primary goals of the mission. The knowledge gathered will pave the path for future developments in space exploration, better our understanding of life on Earth, and inspire the creation of novel technologies that benefit mankind as a whole. By pushing the boundaries of scientific discovery in space, Crew-9 is helping to build a brighter future for all.

Chapter 5

Beyond Low Earth Orbit: Future of Space Exploration

Venturing beyond the comfortable embrace of Low Earth Orbit (LEO) is the next major leap for humanity in space exploration. This cosmic frontier holds great promise for scientific discovery, resource use, and maybe, one day, even human residence. Here's a taste at the intriguing possibilities that lay beyond LEO:

Destinations:

The Moon: Our cosmic neighbor, the Moon, is a natural stepping stone for greater space exploration. Future missions aim to create a persistent human presence on the lunar surface, potentially erecting scientific research centers or even lunar habitats. The Moon's resources, such water ice trapped in polar craters, might be employed for life support and possibly fuel production.

Near-Earth Asteroids (NEAs): These rocky space travelers provide crucial insights to the development of our solar system and perhaps harbor resources like rare metals or water ice. Future missions could involve asteroid exploration, resource exploitation, or even deflection operations to safeguard Earth from impending strikes.

Mars: The Red Planet has enthralled humanity for decades, and robotic expeditions have revealed a world with a remarkably Earth-like past. The Martian dream is to send people on a long-duration mission to create a colony, research the potential for past or current life, and unravel the secrets this distant world holds.

Beyond our Solar System: For the most ambitious minds, the ultimate destination lies outside our solar system. Interstellar travel, however still in the realm of science fiction, is a notion that continues to be investigated. Future improvements in propulsion technology could one day allow humanity to send probes, or possibly crewed expeditions, to study exoplanets in search of life or livable worlds.

Challenges and Technologies:

The path beyond LEO is laden with problems. The harsh environment of space, with its unforgiving radiation and microgravity, needs modern technologies for:

Life Support Systems: Closed-loop systems capable of recycling air, water, and waste will be vital for sustaining human life on long-duration missions.
Propulsion Systems: Powerful and efficient propulsion technologies are needed to cross the huge distances involved in deep space travel. Concepts like nuclear thermal or fusion propulsion are being studied.
Radiation Shielding: Protecting astronauts from dangerous radiation exposure during deep space missions is a crucial topic. Advanced shielding materials and mission planning procedures will be needed.

Habitat Design: Creating self-sustaining habitats that can resist the harsh space environment and provide a comfortable living and working area for humans is a challenging engineering feat.

International Collaboration:

The tremendous undertaking of venturing beyond LEO demands a global effort. International space agencies including NASA, ESA (European Space Agency), JAXA (Japan Aerospace Exploration Agency), and Roscosmos (Russia) are already collaborating on different deep space exploration programs. This united endeavor utilizes experience, resources, and develops a spirit of worldwide cooperation in pushing the boundaries of human space exploration.

The Future Beckons:

Venturing beyond LEO symbolizes a major stride for humanity. It's a monument to our unquenchable curiosity, our need to explore the unknown, and our potential to push the boundaries of scientific and technological growth. The future of space exploration is replete with possibilities, and the Crew-9 mission, albeit constrained to LEO for now, is a stepping stone on this thrilling voyage of discovery. As we journey further into the cosmos, we unleash the possibilities for scientific discoveries, resource harvesting, and possibly one day, fulfilling the dream of becoming a multi-planetary species.

Discussion on the significance of the ISS as a testbed for long-duration spaceflight

The International Space Station (ISS) serves as an excellent testbed for long-duration spaceflight, playing a key role in paving the way for future trips to the Moon, Mars, and beyond. Here's a breakdown of its significance:

Simulating Deep Space Conditions:

Microgravity Environment: Living and working onboard the ISS exposes humans to microgravity for lengthy durations, imitating the conditions experienced during deep space travel. This permits researchers to explore the physiological impacts of microgravity on the human body, including bone loss, muscular atrophy, and cardiovascular deconditioning. By identifying these obstacles, scientists might create countermeasures, such exercise programs and nutritional adjustments, to lessen their influence on astronaut health during long-duration missions.

Closed-Loop Systems: The ISS is a tiny replica of a closed-loop life support system. It recycles air, water, and trash to a certain extent, giving a platform to test and enhance technology needed for long-duration missions reaching far beyond Earth. These systems will be vital for sustaining human life on future spacecraft flying to distant places.

Psychological and Social Aspects:

Crew Dynamics and Teamwork: The multinational crews aboard the ISS contain individuals from varied backgrounds working and living in close quarters for extended durations. This environment provides unique insights into crew psychology, team dynamics, and the importance of good leadership in isolated and confined spaces. The lessons learnt from these interactions are vital for selecting compatible crew members and sustaining a good and productive attitude on long-duration missions.

Technological Advancements:

Testing New Technologies: The ISS serves as a platform to test and refine new technologies necessary for deep space exploration. This includes advanced life support systems, radiation shielding materials, docking procedures for future spacecraft, and technologies for in-situ resource utilization (ISRU), which involves using resources available on celestial bodies like the Moon or Mars to produce essential elements like water or oxygen.

Developing Operational Procedures: The ISS allows astronauts and mission control teams to develop and rehearse operational procedures for various circumstances that might be faced on deep space missions. This comprises emergency response protocols, spacewalk procedures, and maintenance routines for complicated spacecraft systems.

A Stepping Stone to the Future:

The knowledge and experience obtained from the ISS are crucial for planning and performing future deep space missions. It bridges the gap between short-duration spaceflight missions and the grandiose ambitions of pushing farther into the solar system.

Beyond the Benefits:

The ISS is not merely a technical wonder; it's a symbol of international partnership in space exploration. Multiple space agencies have worked together to develop and run this orbiting laboratory, creating a spirit of cooperation and common scientific goals. The success of the ISS lays the path for future collaborative efforts as humanity travels deeper into space.

In conclusion, the International Space Station serves as a vital testbed for long-duration spaceflight. It allows us to investigate the impacts of space on the human body and psyche, enhance the technologies needed for deep space exploration, and establish the operational procedures necessary for venturing further into the cosmos. The ISS is a stepping stone in humanity's great voyage of space exploration, and the knowledge gained from this orbiting outpost will continue to propel us ahead for years to come.

Exploration of commercial opportunities in low Earth orbit and NASA's Artemis campaign for lunar exploration

Low Earth Orbit (LEO) is fast developing into a bustling economic hub, giving tremendous potential for private firms. Here's a breakdown of some significant areas:

Satellite Communications: LEO constellations like SpaceX's Starlink and OneWeb are transforming global internet access, giving high-speed broadband connectivity to remote and underdeveloped locations. This market is predicted to continue developing significantly, with potential for organizations in:
Satellite Manufacturing: Building reliable and cost-effective communication satellites.
Launch Services: Providing launch vehicles to carry these satellites into space.

Ground Station Infrastructure: Developing and operating the network of ground stations that connect with orbiting satellites.

Earth Observation: Private corporations are increasingly using LEO satellites for Earth observation purposes, delivering data on weather trends, agriculture, resource exploration, and disaster monitoring. This presents opportunity for:
Satellite Imagery & Data Analysis: Companies specializing in processing and interpreting satellite data to provide important insights for many sectors.
Constellation Development: Building and running specialized satellite constellations for specific Earth observation uses.
Data Delivery & Services: Developing platforms and services that deliver Earth observation data to clients in user-friendly formats.

Space Tourism: Suborbital and orbital space tourism ventures are growing, allowing ordinary persons to experience the thrill of spaceflight. This market contains potential for:

Spacecraft Development: Designing and manufacturing reusable spacecraft for suborbital and orbital tourist trips.

Space Training & Experiences: Companies offering training programs and arranging space tourism experiences.

Space Hospitality: Developing and running space hotels or space stations for extended space stays.

In-Space industrial: Microgravity in LEO gives new prospects for industrial processes. Companies are exploring the creation of:

Novel Materials: Materials with better qualities that can only be manufactured in microgravity.

Pharmaceuticals: Developing new medications and cures that benefit from microgravity circumstances.

Electronics Manufacturing: Production of specialized electronics that can demand microgravity conditions.

The Artemis Program: Public-Private Partnership on the Moon

NASA's Artemis program intends to return humans to the Moon and establish a sustained lunar settlement. This enormous initiative has significant commercial prospects for private companies:

Lunar Landers and Rovers: Developing and operating lunar landers for crewed and cargo missions to the lunar surface. Companies can also design lunar rovers for exploration and resource transportation.

Lunar Infrastructure: Building and deploying dwellings, communication relays, and resource extraction systems on the Moon will require private sector engagement in design, building, and operation.

In-Situ Resource Utilization (ISRU): Companies can develop methods to harvest water ice from lunar craters and turn it into usable fuel or drinking water, a key resource for a sustainable lunar presence.

Lunar Logistics: Providing transportation and distribution services between Earth orbit, lunar orbit, and the lunar surface will be vital. Companies can construct reusable spacecraft and lunar landers for efficient cargo delivery.

Benefits of Public-Private Partnership:

Reduced Costs: Leveraging private sector expertise and investment can help NASA achieve its lunar exploration goals more cost-effectively.
Technological Innovation: Collaboration between public and commercial sectors can speed the development of new technologies needed for sustainable lunar exploration.
A Thriving Lunar Economy: A public-private collaboration may support the creation of a thriving lunar economy, producing new employment and opportunities on Earth and in space.

Challenges and Considerations:

Regulation and Legal Frameworks: Clearly defined regulations and legal frameworks are needed to manage business activity on the Moon and promote fair competition and responsible resource utilization.

Technical Challenges: Developing the technologies needed for sustained lunar colonization and resource utilization is a complicated task that requires continual investment and innovation.

Risk Management: Space exploration includes inherent hazards. Public-private partnerships need to establish explicit risk management policies to preserve investments and secure the safety of staff.

Conclusion

As we reach the end of our investigation of the Crew-9 mission and its relevance in the grand narrative of space exploration, a plethora of thoughts and emotions come to the forefront. It's a story replete with scientific mystery, technical splendor, and the unwavering human spirit of adventure. The Crew-9 mission serves as a microcosm of humanity's ever-expanding reach into the cosmos, and the knowledge acquired from this attempt will definitely propel us further on this thrilling journey.

At the heart of the Crew-9 expedition lie the scientific experiments, precisely intended to unveil the secrets of life and the universe itself. By examining how living creatures adapt to the microgravity environment, researchers seek to pave the road for eventual space colonization. Understanding the physiological changes astronauts encounter during spaceflight is vital for creating countermeasures to sustain their health throughout long-duration missions, a critical step towards reaching distant destinations like Mars. The Crew-9 mission also contributes to improvements in physical and material research, pushing the boundaries of innovation with tests on fluid dynamics and material testing in space. Furthermore, significant data obtained through Earth observation research will contribute to environmental monitoring efforts and sustainable development methods on our home planet. The ripple effects of these scientific initiatives transcend far beyond the immediate goals of the mission, establishing a spirit of scientific curiosity that benefits all of humanity.

The success of the Crew-9 mission rests on the potent combo – the Dragon spacecraft and the Falcon 9 rocket. The reusability of these vehicles offers a big leap forward in cost-effective space travel, making deep space exploration a more achievable goal. The meticulous design of the Dragon capsule prioritizes human safety, while the strong thrust of the Falcon 9 ensures a successful launch into the celestial vastness. Witnessing the complicated dance of the docking procedures between the Crew Dragon and the ISS is a monument to the incredible precision and engineering prowess that underpins current space operations.

However, the story of space travel goes beyond the mechanical wonders and scientific discoveries. It is a novel weaved with the strands of human grit and perseverance. The astronauts who embark on these risky trips demonstrate incredible bravery and dedication. They train for years, straining their physical and mental limitations to prepare for the rigors of spaceflight. Living in a tight place for extended periods, isolated from loved ones, demands amazing resilience and teamwork. The Crew-9 astronauts symbolize the best of humanity, venturing into the unknown to increase our knowledge and inspire future generations.

The relevance of the International Space Station (ISS) as a testbed for long-duration spaceflight cannot be emphasized. This orbiting laboratory allows researchers to examine the effects of microgravity on the human body and mind, enhance technology needed for deep space travel, and establish the operational procedures necessary for venturing further into the cosmos. The multinational crews onboard the ISS serve as a model for international collaboration in space research, creating a spirit of cooperation and shared scientific goals. The success of the ISS lays the path for future collaborative efforts as humanity travels deeper into space.

Looking beyond Low Earth Orbit, the future of space exploration promises a world rich with possibilities. The Moon, our celestial neighbor, appeals as a stepping stone for deeper space exploration. Future missions aim to create a persistent human presence on the lunar surface, potentially erecting scientific research centers or even lunar habitats. Near-Earth Asteroids (NEAs) contain vital clues to the creation of our solar system and potentially harbor resources for future study and exploitation. Mars, the Red

Planet, continues to grab our imagination, and robotic missions have revealed a world with a remarkably Earth-like past. The Martian dream – to send humans on a long-duration journey to create a colony and unveil the secrets of this strange world – is a monument to humanity's insatiable curiosity and unwavering spirit of exploration.

Venturing beyond our solar system, the concept of interstellar travel, albeit still in the realm of science fiction, continues to fascinate and intrigue. Future improvements in propulsion technology could one day allow humanity to send probes, or possibly crewed expeditions, to study exoplanets in search of life or livable worlds. The expanse of the universe beckons, and the human spirit of exploration has no limitations.

The route towards a future when mankind has a permanent presence beyond Earth is strewn with hurdles. The harsh environment of space, with its unforgiving radiation and microgravity, needs modern technologies for life support systems, propulsion systems, radiation shielding, and habitat architecture. These technical difficulties need continual investment in research and development, pushing the boundaries of scientific innovation. Furthermore, the ethical aspects of space exploration, such as planetary conservation and the potential for resource exploitation, require careful consideration and international cooperation to enable the responsible and sustainable exploration of the universe.

www.ingramcontent.com/pod-product-compliance
Lightning Source LLC
Chambersburg PA
CBHW070347230526
45471CB00006B/2447